The Near Decline Of Physics Due to its Undressed Terms

Austin P. Torney

HIGGS BOSON

PROMETHEUS UNBOUND

STRANGE QUARK

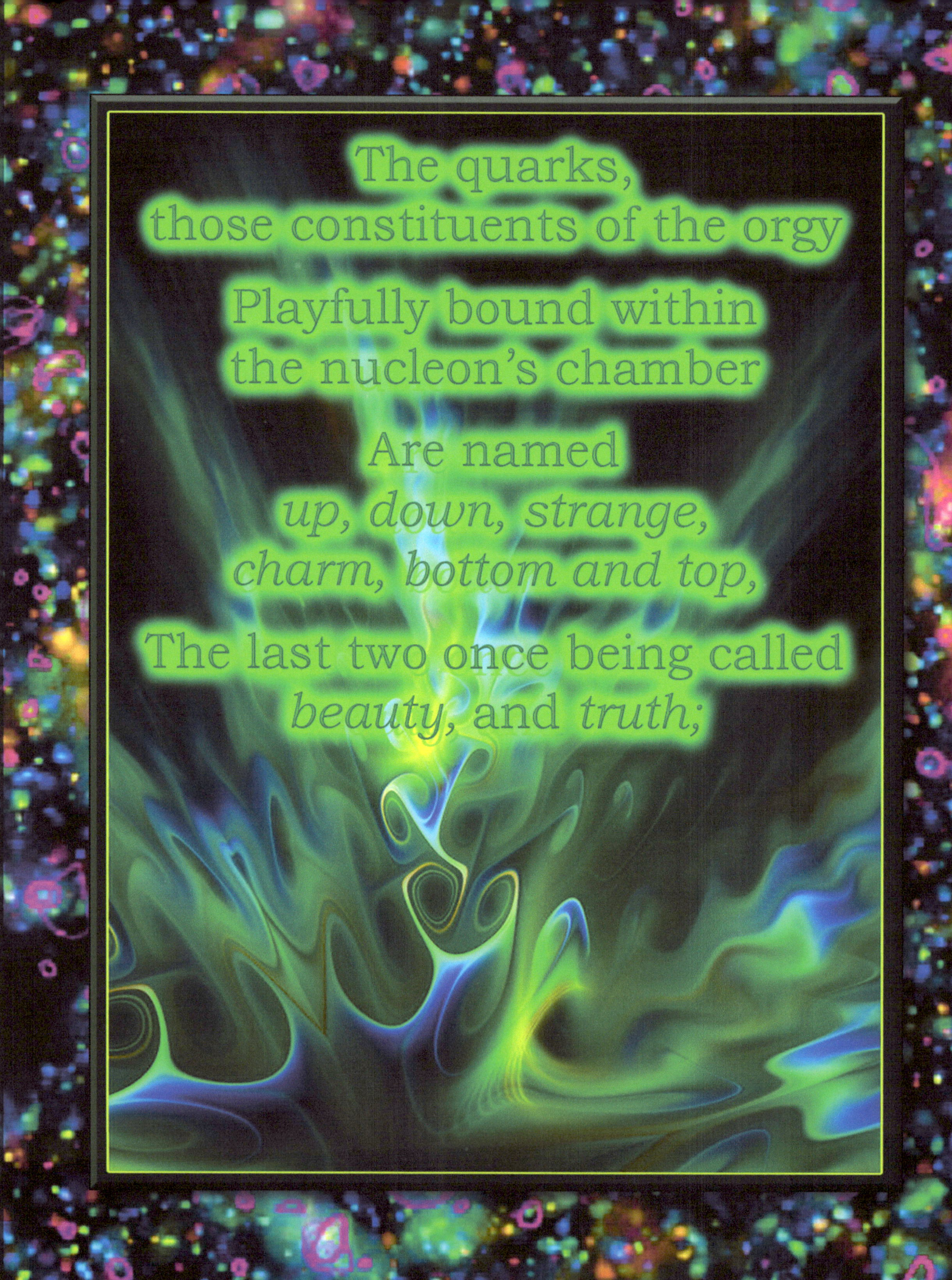

The quarks,
those constituents of the orgy

Playfully bound within
the nucleon's chamber

Are named
*up, down, strange,
charm, bottom and top,*

The last two once being called
beauty, and *truth;*

However, when just one of a type was contained
It became referred to, say, as a naked beauty,
And thus nude tops & bottoms
their charms revealed—
To ever be in closeness binding, and bonding.

So, they even tried just *u, d, s, c, b,* and *t*
To prevent some ultimate collapse of physics,
But the truth of the flavors beneath the veils
Remained as the sheerest vision preferred.

So, we have these vibrant dancing ladies:
The naked heavyweight top, charming up,
And, down, the strange beauty of the raw truth,
With a bare bottom just around and behind.

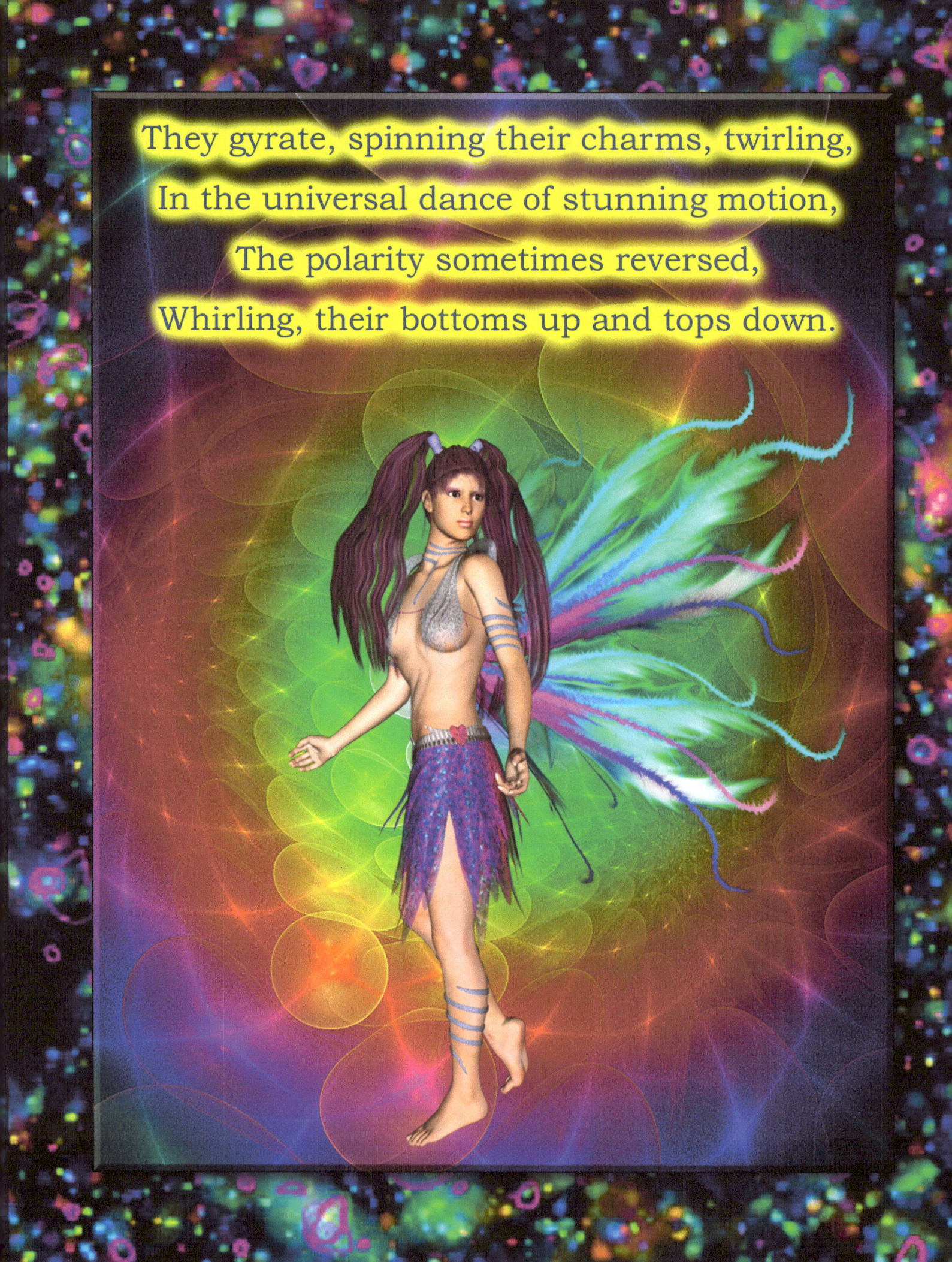

They gyrate, spinning their charms, twirling,
In the universal dance of stunning motion,
The polarity sometimes reversed,
Whirling, their bottoms up and tops down.

Gluons are the bees of the flower beds,
Carrying pollen back and forth to bond
The many relationships that make
This loved world go 'round as reality.

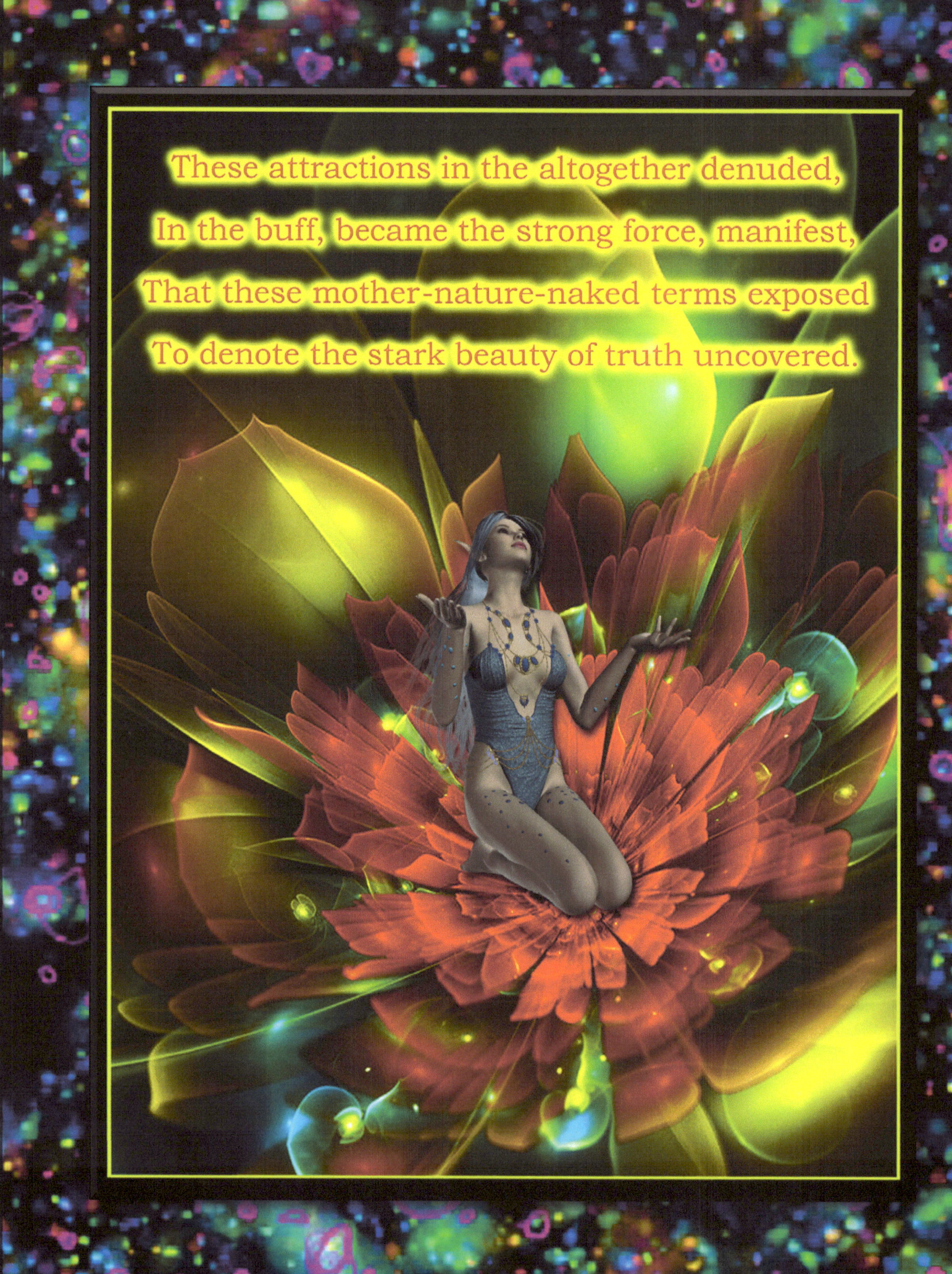

These attractions in the altogether denuded,
In the buff, became the strong force, manifest,
That these mother-nature-naked terms exposed
To denote the stark beauty of truth uncovered.

THE ENTRANCING DANCING

They were all dancing within love's treasure vault
Within the framework of the broadening thought,
The lights pulsing and the waves reverberating,
Where the good times had become everlasting.

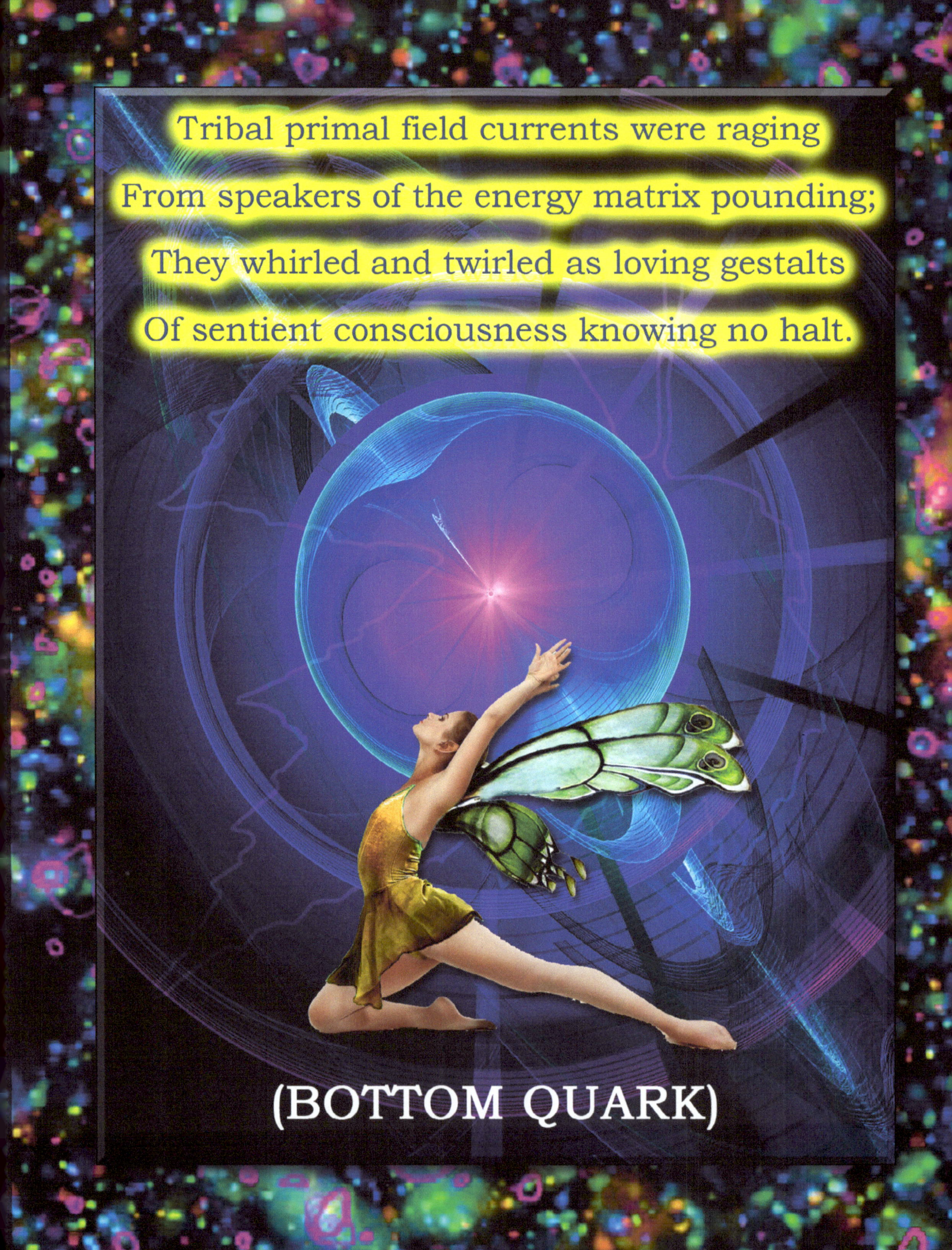

There were rhythms of constant contraction
And expansions of bosom-energy projections
Converted to scalar waves of blinking attraction,
As fission and fusion beckoned the connections...

Ever forming in this Omni-sound emporium,
Where tone waves vibrated in waves of creation.

"THREE QUARKS FOR MUSTER MARK"

Naked quarks would really love
to go wild and dance,

But there's only a finite amount
of energy and chance;

So, they would spiral out of control,
Having quite a blast!

Such, they've been confined
within the proton—
To last.

No, for the quantum censor protects the charm show,
Their strange beauty and flavor bound up and down,
For the proton is much immune to disturbance around.

CHARMS

A new kind of microscope
That works via
gravitational waves
Has revealed
the actual interior
Of a quark for the first time.

The charming beauty
Of the ultimate truth
Is that ladies are
In charge of the universe!

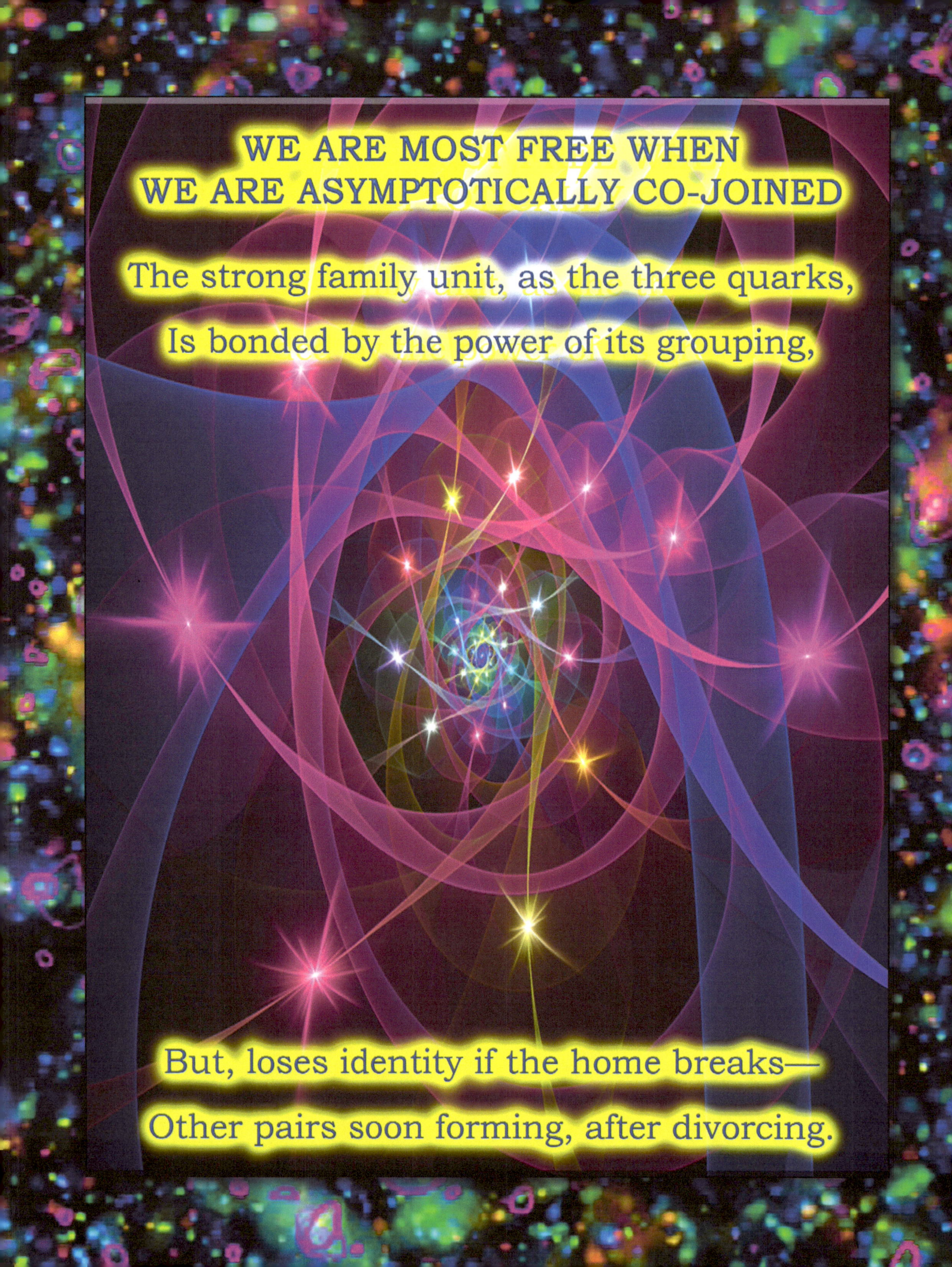

WE ARE MOST FREE WHEN WE ARE ASYMPTOTICALLY CO-JOINED

The strong family unit, as the three quarks,

Is bonded by the power of its grouping,

But, loses identity if the home breaks—

Other pairs soon forming, after divorcing.

Or comes the prison of solitude,
Chained to isolation with fortitude,
Floating, lost, without effects of affects,
Losing the identity conferred by others.

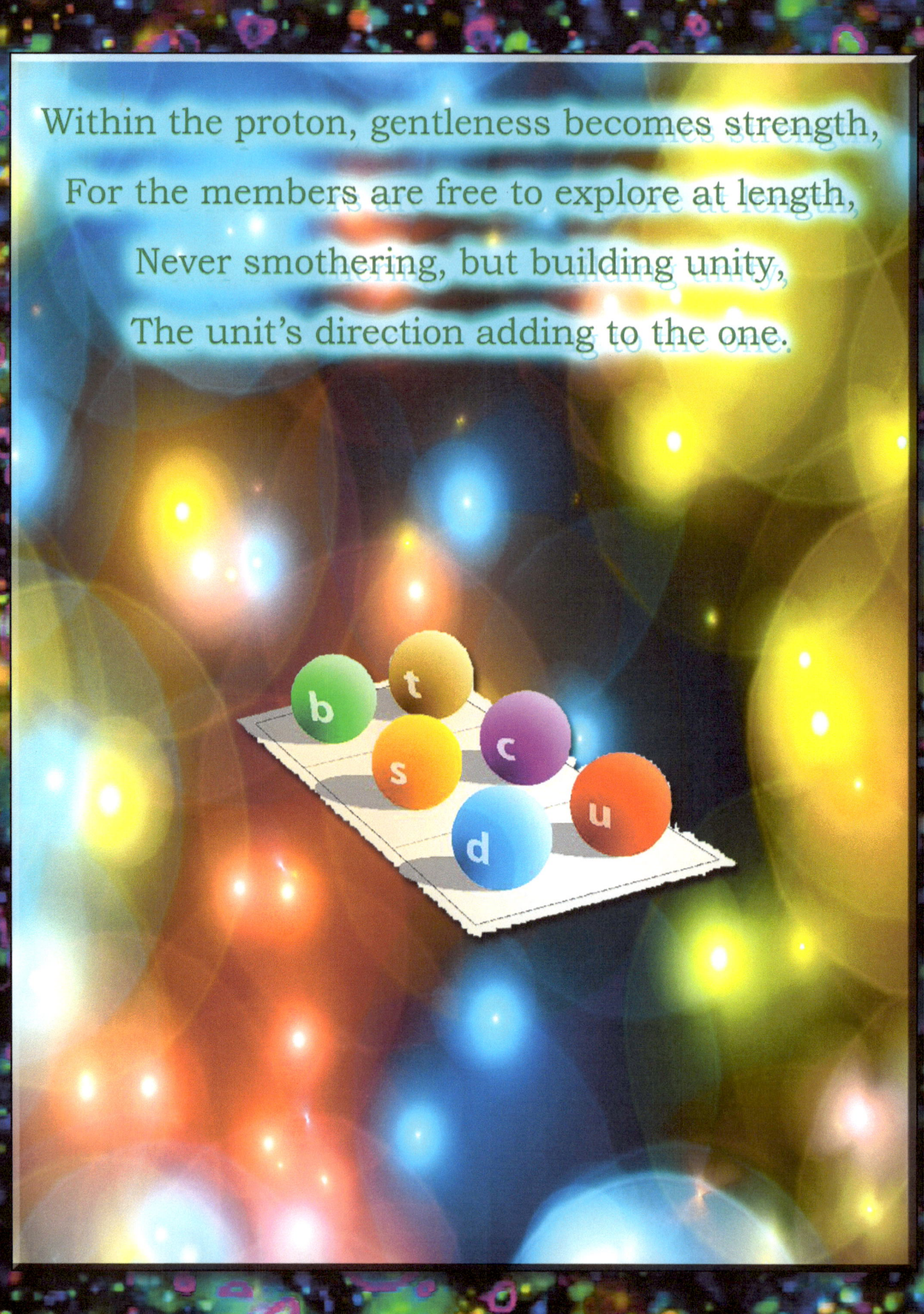

Within the proton, gentleness becomes strength,
For the members are free to explore at length,
Never smothering, but building unity,
The unit's direction adding to the one.

Identity is not lost in the co-joining—
True loves don't crowd the hearts of each other,
But, rather, look outward, in the same direction,
Close, joined, but not blocking the others section.

THE PROTON—LIGHTNING IN A BOTTLE

A proton is made of
three quarks, true,

But quarks are so small

That they only make up

Two percent or so

Of the proton's total mass.

proton
anti-proton
electron
positron
hydrogen
anti-hydrogen

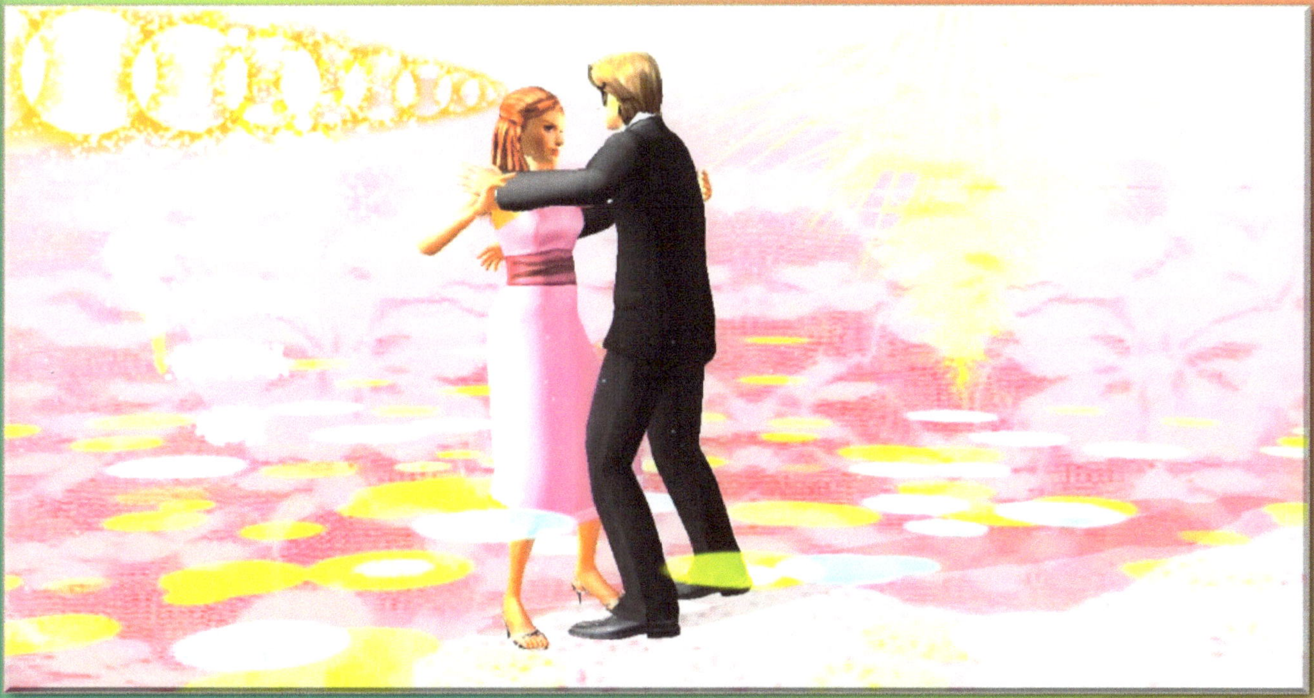

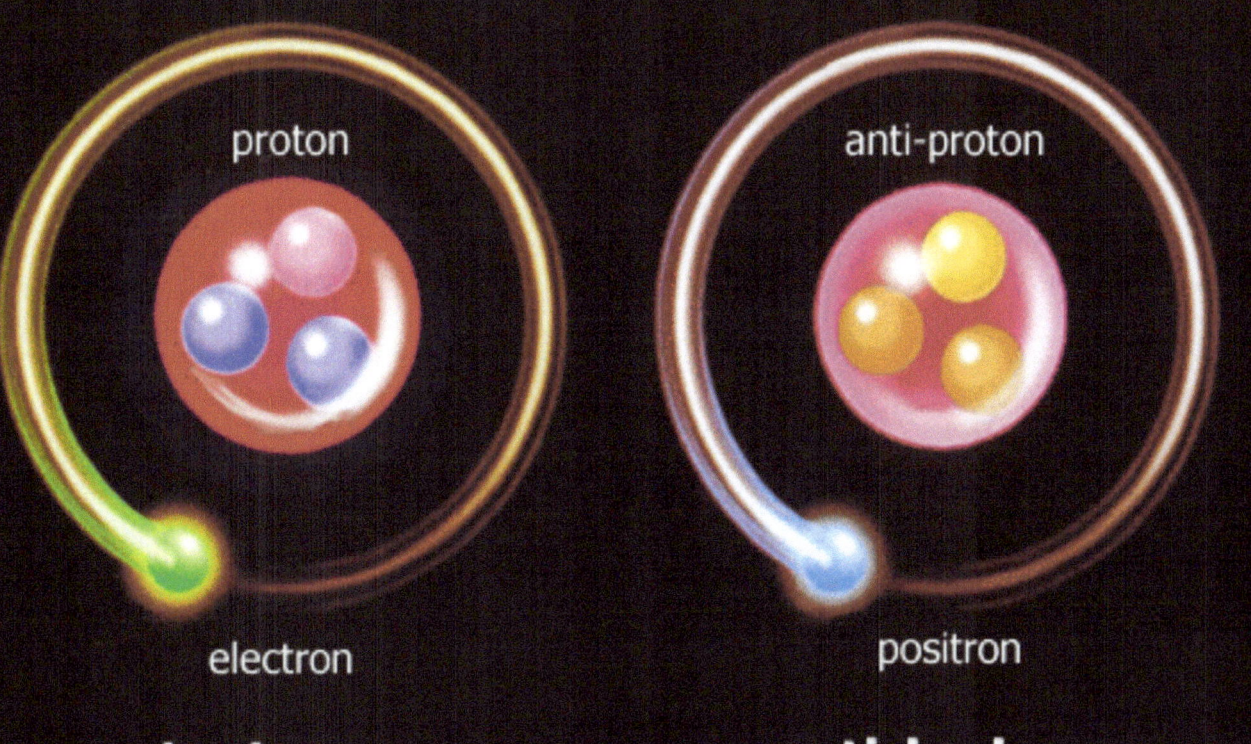

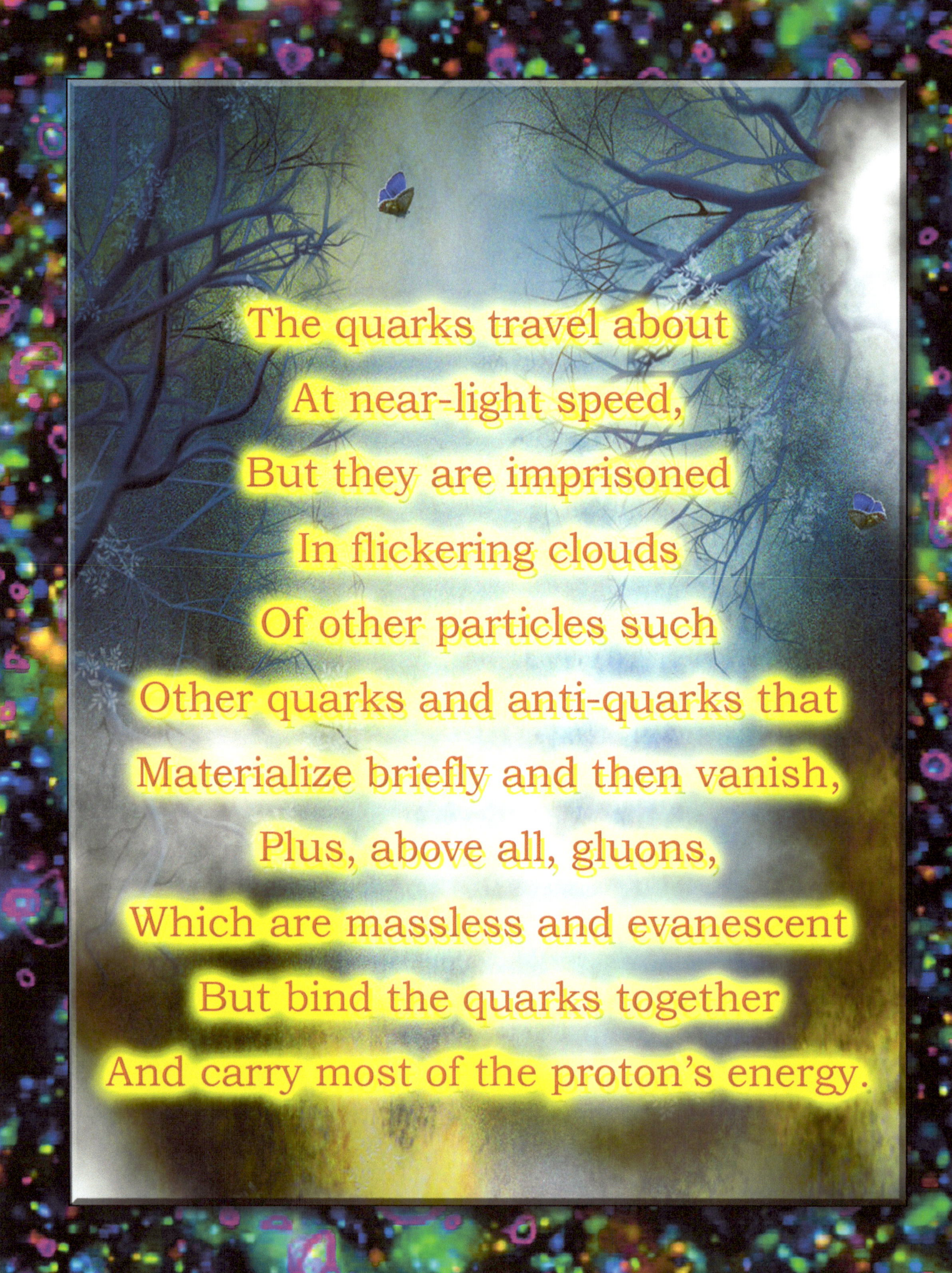

Such, it would be accurate to say

That protons are made of gluons

Rather than quarks.

Experiments have proved

Wilczek's descriptions correct.

Every proton now around

Congealed from

A quark-gluon plasma soup

That existed for a microsecond

After the Big Bang.

In QED,
Which is about interactions
Of electromagnetic fields,
A screen made of photons
And short-lived
electron-positron pairs
Partially cancels
the electric fields,
Although, on the inside,
The fields get much stronger;
However, in QCD,
The opposite happens,
Called antiscreening—
The strong force
between the quarks
Gets weaker as the quarks
Get closer together.

So the quarks seem to be
"Asymptotically free",
Moving about as if there
Were no force between them at all
But this freedom is an illusion,
For, as the distance
between them increases,
So does the strong force.

It is that the proton has parts,
But it cannot be taken apart!

The proton is mostly glue,
A kind of a super glue.

One number remains
constant in protons
As their virtual quarks
and anti-quarks
Come and go in
their uncountable swarms:

Three.

Histoire D'amour
Love Story

Un homme tombe en amour avec ses yeux,
A man falls in love through his eyes,

Une femme travers ses oreilles;
A woman through her ears;

Plus tard, il renverse …
Later it reverses…

Une femme prend note de tout ce qui sera fait,
A woman notes everything to be done,

Mais l'homme ne connaot pas le voir un.
But the man does not hear the seeing one.

Mais il ya encore de l'espoir …
But there is still hope…

Comme dans le mariage
As in the marriage

De la femme aveugle
Of the blind lady

Pour l'homme sourd.
To the deaf man.

www.ingramcontent.com/pod-product-compliance
Lightning Source LLC
Chambersburg PA
CBHW050741180526
45159CB00003B/1302